Experientia Supplementum 29

Synthesis of the Tripeptide L-Trp-L-Ser-L-Glu

Comparison of its biological activity
with that of the Delta-Sleep-Inducing-Peptide (DSIP)

By Vincent M. Monnier

MD, Dipl. Chem., Geneva

Birkhäuser Verlag

Basel und Stuttgart

Experientia Supplementum 29

Synthesis of the tripeptide L-Trp-L-Ser-L-Glu

Comparison of its biological activity with that of the *D*elta-*S*leep-*I*nducing-*P*eptide (DSIP)

Vincent M. Monnier
Dr. med. and Dipl. Chem.

1977 Birkhäuser Verlag, Basel and Stuttgart

CIP-Kurztitelaufnahme der Deutschen Bibliothek

Monnier, Vincent M.
Synthesis of the tripeptide L-Trp-L-Ser-L-Glu:
comparison of its biolog. activity with that of
the Delta-Sleep-Inducing-Peptide (DSIP). - 1.
Aufl. - Basel, Stuttgart: Birkhäuser, 1977.
(Experientia: Suppl.; 29)
ISBN 3-7643-0915-6

ISBN 3-7643-0915-6

Table of contents

Preface

The present work was done, as concerns the chemical-synthetical part, at the Institute for Organic Chemistry of the University Basel under the supervision of Prof. Dr. M. Brenner, as concerns the biological part, at the Laboratory for Sleep Research of the Physiological Institute of the University under the supervision of Prof. Dr. M. Monnier.
I should like to express my gratitude to PD Dr. G.A. Schoenenberger who suggested the subject of this thesis and provided precious advice for its realization. I am very much indebted to Prof. Dr. M. Brenner who directed the synthetic work of the tripeptide. I am also grateful to my father Prof. Dr. M. Monnier and his coworkers L. Dudler and R. Gächter who tested the activity of the synthetical tripeptide. I also thank M. Grogg, Bachem, Liestal, for his help as concerns hydrogenolysis reactions and E. von Arx for amino acid analysis at the Chromatographie-Labor, Pharma-Forschung, CIBA-GEIGY AG, Basel.

Abbreviations

The abbreviations for amino acid derivatives, protecting groups and further helper groups are employed according to the recommendations of the IUPAC-IUB Commission on Biochemical Nomenclature which are reported in Biochem. J. *126,* 773–780 (1972). Amino acids without symbol for the direction of rotation have the L-configuration.

Z = benzyloxycarbonyl (earlier carbobenzoxy-)
BOC = tert.-butyloxycarbonyl-
TFA = trifluoroacetyl-, trifluoroacetate
BZL = benzyl-
TMS = tetramethyl silane

Symbols which stand *in front* of the abbreviation for the amino acid residue mean substitution on the α-amino group, symbols *behind* substitution on the α-carboxyl group. Substituents for a third functional group are given in brackets.

For example:

H-Ser-OH for L-serin
BOC-Ser(BZL)-OH for N-α-tert.-butyloxycarbonyl-O-benzyl-L-serin

I Introduction

The concept that sleep and waking are controlled by some chemical change dates back to the early 1900s: LEGENDRE and PIÉRON [1] showed that infusion into subarachnoidal spaces of liquor cerebrospinalis, total brain extract and serum from a ten-day sleep deprived dog to another dog induced a behavioral sleep in the latter. They suggested the existence of 'hypnotoxins' responsible for that sleep. In 1927, HESS [2] induced sleep in the cat by electrical stimulation of the ventral thalamus, near to the commissura interthalamica. His criteria of normal sleep in the cat were suppression of spontaneus contact with environment while intact arousal reactivity to external auditory, olfactory and tactile stimuli remained, constriction of the pupil, closure of the nictating membrane and of the eyelids, typical sleep posture.

In 1963, MONNIER sen., KOLLER and GRABER [3] showed that a natural substance, produced in donor rabbits kept asleep by electrical stimulation of the thalamic area of HESS, and transmitted by crossed blood circulation to a recipient rabbit, induced sleep in the latter. Sleep in rabbits restrained in a hammock, was characterized on the one hand by visceral effects (miosis, partial closure of the eyelids, decreased respiration rate) and intact arousal reaction to external stimuli, on the other hand by electroencephalographic effects (EEG): initial appearance of spindles (14 Hz) at the beginning of sleep in the motor, sensomotor and limbic cortex, followed by low delta frequencies typical of deep sleep. These EEG phenomena are symptomatic of orthodox 'Slow Wave Sleep' (SWS) in contrast to 'paradoxical' sleep (REM-Sleep) [4, 5]. Through extracorporal dialysis of venous blood from the sagittal sinus during thalamic induced sleep, it was possible to obtain [6] from rabbit donors a hypnogenic hemodialysate. This elicited delta-sleep in waking recipient rabbits 20 min after intravenous injection.

These investigations suggested a humoral transmission of sleep and initiated further investigations in order to extract possible natural hypnogenic factors from various body fluids.

Today there are mainly three groups working on transmission of sleep by natural substances. PAPPENHEIMER et al. [7] reported the fractionation by ultrafiltration through molecular sieves membranes of a substance, Sleep-Promoting Factor S, extracted from cerebrospinal fluid and from brains of sleep deprived goats. Sleep promoting activity was assayed by decrease in nocturnal locomotor activity of rats and by increased duration and amplitude of slow wave cortical EEG activity in rabbits. The substance was subsequently identified as an oligopeptide containing Asp, Thre, Gly and possibly Glu/Gln or Ser as shown by amino acid analysis. NAGASAKI et al. [8] reported the partial isolation of a low molecular weight material extracted from brain stem of sleep deprived rats, which caused reduction of nocturnal locomotor activity and increased slow wave EEG in the recipients. The team of MONNIER sen. and SCHOENENBERGER extended the previous investigations on hypnogenic hemodialysate and isolated in a first step a 'Sleep Factor Delta' [9] which induced delta-sleep upon intraventricular infusion in the rabbit. In a second step, this Factor Delta was purified, characterized and identified [10] as a peptide (molecular weight 800–900 daltons), now called Delta-Sleep-Inducing-Peptide (DSIP), with the following amino acid composition: 1 Ser, 1 Glu, 3 Gly, 2 Ala, 1 Asp and tryptophan as N-terminal amino acid [11]. The amino acid sequence and synthesis were reported in 1976 and recently published [25]. The purpose of the present work was to synthesize a tripeptide with the same N- and C-terminal amino acids in order to test the specificity of the natural Delta-Sleep-Inducing-Peptide by contrast to other peptides of identical amino acid composition. Therefore, tryptophan as N-terminal, glutamic acid as C-terminal amino acid residue, and serine since it is found in the 'active site' of several peptides and enzymes, were chosen. We report here the synthesis of L-tryptophyl-L-seryl-L-glutamic acid and the results of the biological tests.

II Schema of synthesis

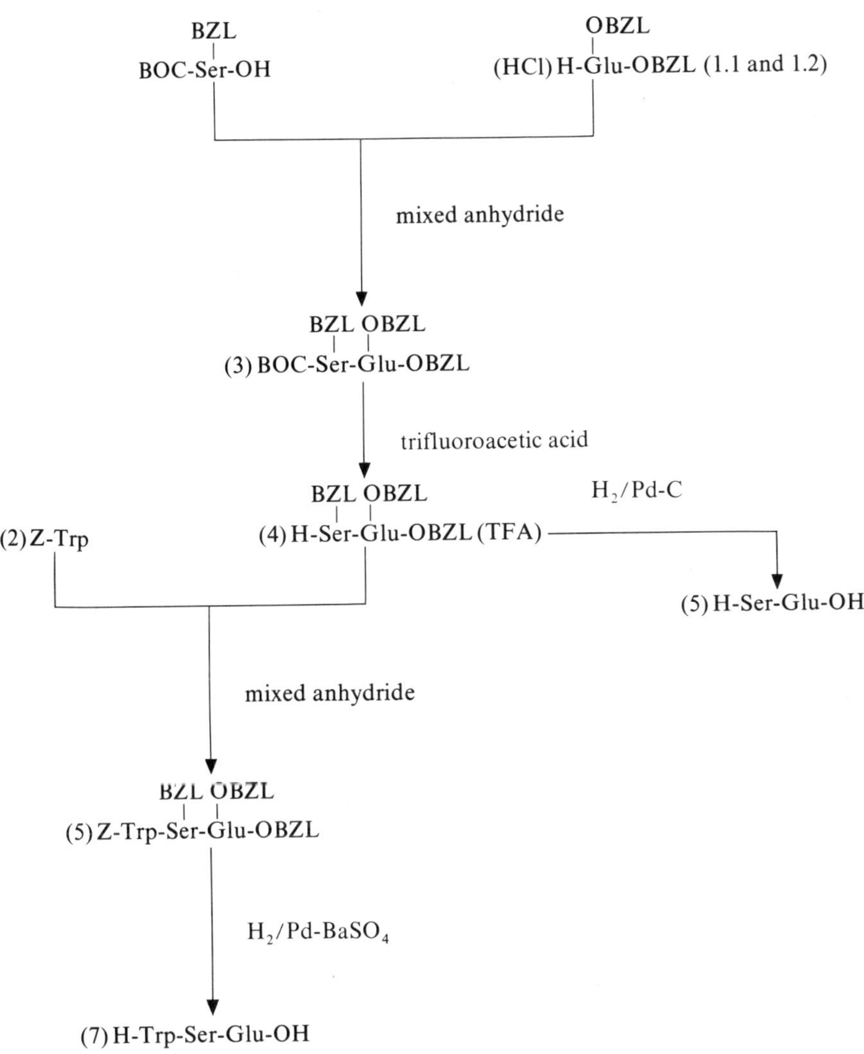

III Material and methods

Amino acids and derivatives: L-glutamic acid and L-tryptophan (grade puriss.) were supplied by Fluka, Buchs, O-benzyl-L-serin was from Bachem, Liestal. Solvents and reactives of the quality 'for synthesis' or 'pure' were from Siegfried, Zollikofen, and Fluka, catalyst for hydrogenation (10% palladium on charcoal) was from Johnson-Matthey, London, and (5% palladium on barium sulfate) from Engelhart, Hannover.

Synthesis of the compounds I–VII

The physico-chemical data of the synthesized compounds will be given under results.

1.1 *H-Glu(OBZL)-OBZL benzosulfonate and hydrochloride*

Procedure of SHIELDS et al. [12].

(a) 14.7 g (0.1 mol) L-glutamic acid, 18.5 g (0.105 mol) benzene sulfonic acid monohydrate, 70 ml (6 time excess) benzyl alcohol and 200 ml carbon tetrachloride were refluxed for 15 hrs. The reflux condensate was continuously passed through a Soxhlet extraction apparatus containing silica gel. 50 ml carbon tetrachloride were added to the warm reaction mixture which was subsequently stored in the cold overnight. The crude product was recrystallized from methanol and diethyl ether and the colorless crystals were dried over phosphorus pentoxide in vacuo at room temperature. Yield 39.5 g (81% of the theory).
(b) Conversion of the benzosulfonate to the hydrochloride was achieved by mixing under stirring 20 g dibenzyl ester (1.1a) (0.04 mol) with an equimolar quantity triethylamine (5.8 ml) in 90 ml chloroform until dissolved. The benzosulfonate salt of triethylammonium was precipitated with absolute diethyl ether and removed by filtration. Dry hydrogen was bubbled through the cold filtrate until all the dibenzyl ester hydrochloride was precipitated (about 20 min). Colorless crystals were recrystallized from methanol and diethyl ether. Yield 12.6 g (87% of the theory).

1.2 *H-Glu(OBZL)-OBZL p-toluene sulfonate and hydrochloride*

The esterification procedure described by ZERVAS et al. [13] was used.

(a) 10 g L-glutamic acid (0.068 mol), 50 ml benzyl alcohol, 12.96 g (0.069 mol) *p*-toluene sulfonic acid monohydrate and 50 ml benzene were refluxed for 5 hrs and the water formed (3.6 ml) azeotropic removed. Benzene (70 ml) and dry diethyl ether (100 ml) were added to the reaction mixture, the crystalline diester collected and recrystallized from methanol and ether. Yield 75%.
(b) Conversion to the hydrochloride was accomplished by the method of HILLMAN [14].
16.5 g (0.33 mol) dibenzyl ester *p*-toluene sulfonate (1.2a) were dissolved in 25 ml chloroform and added to 19.2 ml of a 2% ammoniacal chloroform solution (cold chloroform saturated with NH_3 gas). After stirring and cooling for 15 min, the precipitated ammonium tosylate was filtered off and washed with little chloroform. The solvent was removed under reduced pressure at room temperature and the oily residue dissolved in dry diethyl ether (100 ml). Hydrogen chloride was bubbled through the filtrate and the product precipitated. Recrystallization from methanol and diethyl ether. Yield 10.4 g (87% of the theory).

2 *Benzyloxycarbonyl-L-tryptophan*

Prepared by the method of SMITH [15].

To a solution of 5.1 g (0.025 mol) of L-tryptophan in 25 ml NaOH 1 M at 0 °C were added, with cooling and shaking, 4.2 g (0.0247 mol) of carbobenzoxy chloride and 25 ml of NaOH 1 M. 1 hr after the last addition, the solution was acidified to pH 4 with HCl 5 M. The precipitate was filtered off and recrystallized from ethyl acetate and petroleum ether. Yield 7.4 g (87% of the theory).

3 *BOC-Ser(BZL)-Glu(OBZL)-OBZL*

According to the method for mixed anhydride of WÜNSCH and ZWICK [16].

To a solution of 5.7 g (0.0193 mol) BOC-Ser(BZL)-OH in 30 ml dry tetrahydrofuran (THF) and 2.62 ml triethylamine were added 1.85 ml (excess) ethyl chlorocarbonate at $-10\,°C$. After 15 min at $-10\,°C$, a solution of 6.72 g L-glutamic acid dibenzyl ester hydrochloride and 2.62 ml triethylamine (equimolar) in 50 ml THF was added with stirring. After 2 hrs at $-10\,°C$ and 12 hrs at room temperature, the mixture was concentrated at 40 °C and shaked with a large volume (500 ml) of water. A yellow oil was thus percipitated and 'frozen' at 0 °C. The supernatant was decanted from the solid oil and the residue dissolved in warm ethyl acetate (100 ml). This solution was washed 3 times each with 10% citric acid, potassium bicarbonate 1 N and with water, dried over $MgSO_4$ and evacuated at 40 °C. A colorless oil remained which solidified within two days drying at 0.05 torr over P_2O_5 at room temperature. Yield 98%.

4 *Ser (BZL)-Glu(OBZL)-OBZL trifluoroacetate*

11.13 g (0.0184 mol) BOC-dipeptide ester (3) were added to 20 ml cold trifluoroacetic acid and the mixture stirred for 1 hr. The remaining free acid was evacuated as far as possible under vacuum and high vacuum. A yellowish sirup was obtained with 126% yield. The sirup was taken up in 30 ml dry THF and the excess TFA neutralized exactly with 3.39 ml triethylamine. This solution was stored at 4 °C and directly used for subsequent synthesis.

5 *Z-Trp-Ser(BZL)-Glu(OBZL)-OBZL*

Same procedure as for peptide 3.

To a mixture of 2.4 g (7.1 mmol) Z-tryptophan (2) in 15 ml dry tetrahydrofuran and 0.964 ml triethylamine (equimolar) were added 0.67 ml ethyl chlorocarbonate at $-10\,°C$. After 15 min, a mixture of 7.1 mmol dipeptide diester trifluoroacetate (4) and 0.964 ml triethylamine was added with stirring. After 2 hrs at $-10\,°C$ and 12 hrs at room temperature, the solvent was evacuated at 40 °C and the residue taken up in warm acetic acid. The crude product was washed 12 times with 0.3 N HCl, 6 times with $KHCO_3$ 1 N and 4 times with water. On evaculation of the solvent, precipitation of the product began and was achieved with ether. Recrystallization from acetone and petroleum ether. Yield 3.25 g (56%) of colorless amorph crystals.

6 *Ser-Glu*

Hydrogenolysis according to Fölsch [24].

5.0 mmol of the trifluoroacetate (4) in THF solution were concentrated in vacuo. The resulting oil was diluted in cold ethyl acetate, rapidly washed at 0 °C twice with 0.5 N $KHCO_3$, 3 times with water and dried over $MgSO_4$. The free base was taken up in 25 ml tert.-butyl alcohol/water (3:1) and hydrogenolyzed over 0.5 g 10% palladium on charcoal. After 2 hrs, the solution was filtrated over Cellit and evacuated at 40 °C. Colorless needles precipitated in wet methanol and dry diethyl ether. Yield 0.84 g (81%). After drying overnight at 0.05 torr over P_2O_5, the dipeptide took up exactly 1 equivalent water.

7 *Trp-Ser-Glu*

0.8 g Z-tripeptide dibenzyl ester (5) (0.98 mmol) were dissolved in methanol/dimethylformamid (8:2) and hydrogenolyzed over 100 mg 5% Pd-$BaSO_4$. After 5 hrs, the formation of CO_2 has ceased as shown by $Ba(OH)_2$ absorber. After filtration over cellit, a few drops of water in methanol were added and a hygroscopic product was precipitated with diethyl ether. 20 mg of this intermediary product (cf. results and discussion) were dissolved in 25 ml tert.-butyl alcohol/water (3:1) and rehydrogenolyzed according to Grogg, Bachem (Liestal), over 150 mg Pd-$BaSO_4$ during five days and treated as above. Yield 0.015 g of slightly yellowish crystals.

IV Results

The analyses were performed at the Labor für Mikro-Analysen of the Institute for Organic Chemistry Basel. The melting points were determined using a Kofler bench and were corrected. The specific optical rotation was measured with the polarimeter 141 of Perkin-Elmer. Purity control was performed with paper chromatography. Solvents frequently used were:

A: 1-butanol–acetic acid–water, 4:1:1 (R_{fA})

B: Benzone–chloroform–methanol, 10:6:1 (R_{fB})

The spots were revealed with UV, iodine, ninhydrin and Ehrlich reagent for tryptophan.

Compound	No. or ref.	Melting point °C	Optical activity $(\alpha)_D$ Temp.	Conc.	Solvent	Value	TLC R_{fA}	Chrom. R_{fB}	Analysis
H-Glu (OBZL)-OBZL benzosulfonate	1.1a	117–119	24	3.0	EtOH	+ 5.0			
	[12]	115–117		item	96%	+ 8.4			
H-Glu (OBZL)-OBZL hydrochloride	1.1b	101–103	24	3.1	pyridin	+ 15.2			
	[12]	100–102		item					
H-Glu (OBZL)-OBZL p-toluosulfonate	1.2a	140–142	24	2.0	MeOH	+ 7.2			
	[13]	144–145		item		+ 7.6			
	[17]	140	20	2.0	EtOH	+ 7.2			
H-Glu (OBZL)-OBZL hydrochloride	1.2b	99–101	24	2.0	HCl	+ 8.8			
	[13]	100–102		item		+ 9.82			
	[18]	100–102	24	3.0	HCl	+ 9.4			
Z-Trp	2	123–124	23	5.0	1 eq.	− 14.4			
	[15]	126	not reported		NaOH				
D-isomer	[19]	136–137	25	5.0	item	+ 15.5			
BOC-Ser (BZL)-Glu (OBZL)-OBZL	3	46– 47	24	2.0	tetrahydrofuran	+ 2.7	0.90	0.65	$C_{34}H_{40}N_2O_8$ calc. C 67.59% H 6.66% found C 67.53% H 6.67%

Compound	No. or ref.	Melting point °C	Optical activity $(\alpha)_D$ Temp.	Conc.	Solvent	Value	TLC R_{fA}	Chrom. R_{fB}	Analysis
Ser (BZL)-Glu (OBZL)-OBZL trifluoroacetate	4						0.64		
Z-Trp-Ser (BZL)-Glu (OBZL)-OBZL	5	125–126	24	2.0	aceton	– 10.63	0.95 both ninh. and Ehrlich pos.	0.31	$C_{48}H_{48}N_4O_9$ calc. C 70.10% H 5.83% found C 69.88% H 5.86%
Ser-Glu	6	138–139	24	1.75	HCl 1 N	– 7.8	0.13		$C_8H_{14}N_2O_6 \cdot H_2O$ calc. C 38.06% H 6.39% found C 38.06% H 6.91%
	[24]	not rep.	26.5	4.2	HCl 1 N	– 11.4	0.15		
	[20]		25	6.0	HCl 1 N	– 9.4			
Intermediary product Trp-Ser (BZL)-Glu		202–205	22	2.7	DMF/ H_2O 2:1	+ 25.1	0.57		$C_{19}H_{28}N_4O_9 \cdot H_2O$ calc. C 52.03% H 5.52% found C 51.32% H 6.36%
Trp-Ser-Glu	7						0.35 ninh. and Ehrlich pos.		Amino acid analysis: Trp 0.64, Ser 0.64, Glu 1.0

V Discussion

(a) *Chemical part*

In a first series of experiments, glutamic acid was esterified according to the procedure of SHIELDS et al. [12]. The resulting ester was partially racemized, probably due to the alkalis present on glass walls during the long heating period. When the procedure of ZERVAS et al. [13] was used, better values for the optical activity, but a worse yield were obtained.
The reaction of tryptophan with benzyl oxycarbonyl after SMITH [15] yielded Z-Trp in the same range. However, as no data for its optical activity were reported, we had to compare with values for the D-isomer [19].
The coupling reaction to the protected dipeptide was quantitative and yielded an oil which could be easily purified and solidified, however only after drying under high vacuum. The purification of the protected tripeptide from unreacted products was more difficult because of solubility problems arising during the extraction procedure.
The cleavage of the BOC group with trifluoroacetic acid was quantitative, but the excess of TFA could not be evacuated even under high vacuum. Consequently it had to be neutralized with triethylamine in order to avoid acid hydrolysis during storage. For the hydrogenolysis reaction to the free dipeptide, the free base was shaken with bicarbonate in the cold in order to avoid formation of dicetopiperazine. The resulting peptide L-Ser-L-Glu had a smaller optical rotation value than the one reported by FÖLSCH [24] and furthermore contained 1 equivalent water. The hydrogenation of the protected tripeptide with palladium on charcoal in acetic acid yielded a strong rosa colored and TLC sensitive product. This color was also observed and commented upon by WÜNSCH [21] in his chapter 'Tryptophan induced side-reactions': "After hydrogenolytic treatment of N-Z-peptides, especially of strong acidic peptides or in presence of acetic acid, one observes a red-violet coloration which can be abolished or diminished by working in oxygen poor (N_2, argon) atmosphere." The hydrogenation with $Pd/BaSO_4$ under both acidic and neutral conditions yielded a hygroscopic compound with the main spot at $R_{fA} = 0.57$ and a

smaller one at $R_{fA} = 0.35$. The suspicion that not all benzyl protecting groups were cleaved was confirmed through ^{1}H-NMR-spectroscopy [22]: The spectra A and B (Fig. 1) show that the three singletts a, a′ and a″ between 5.0 and 5.2 ppm disappear after hydrogenation, while that of the benzyl ether of serine remains at 4.4 ppm (b). The integral of aromatic protons h, h′ and h″ suggests that 10 instead of 5 not exchangeable protons are present after hydrogenation. On prolonged rehydrogenation, five days over 5% Pd/$BaSO_4$ in tert.-butanol/water, the tripeptide was liberated. Spectrum C no longer exhibits singletts and the aromatic integral corresponds to 5 indole protons in D_2O.

(b) *Biological part*

The bioassays of the tripeptide whose synthesis has been described were carried out in the Sleep Research Laboratory of the Physiological Institute (University of Basel) by L. Dudler and R. Gächter.

Intraventricular infusion test in the rabbit. A technique for infusion of chemical substances into the third ventricle of the rabbit was developed by Monnier sen. and Hatt [23]. A cannula was introduced into the brain of the rabbit, restrained in a hammock, between the third ventricle and the acqueductus of Sylvius. The effects of the infused substance were recorded electrographically with electrodes in the motor cortex (right and left). Concurrently, visceral effects (heart rate = ECG and respiration rate = EPG) were recorded.

Each experiment lasted 95 min and was divided into three periods. After adaptation of the rabbit to its environment (shielded and soundproof cage), the EEG was recorded during a *pre-infusion period* of 20 min. This was followed by a second 20-min period consisting of the intraventricular infusion (3.5 min) and the subsequent recovery period (16.5 min). The pre-infusion period and the following *infusion-recovery period* were used as reference level (40 min altogether). The statistical comparison of the pre-infusion delta value to that of the infusion-recovery period showed that either period of 20 min or both together could be used as reference for the delta activity changes occurring during the following *post-infusion period* (55 min). This third period was considered as the *specifically active period.*

Quantification of the effects. The EEG parameters of activity chosen were

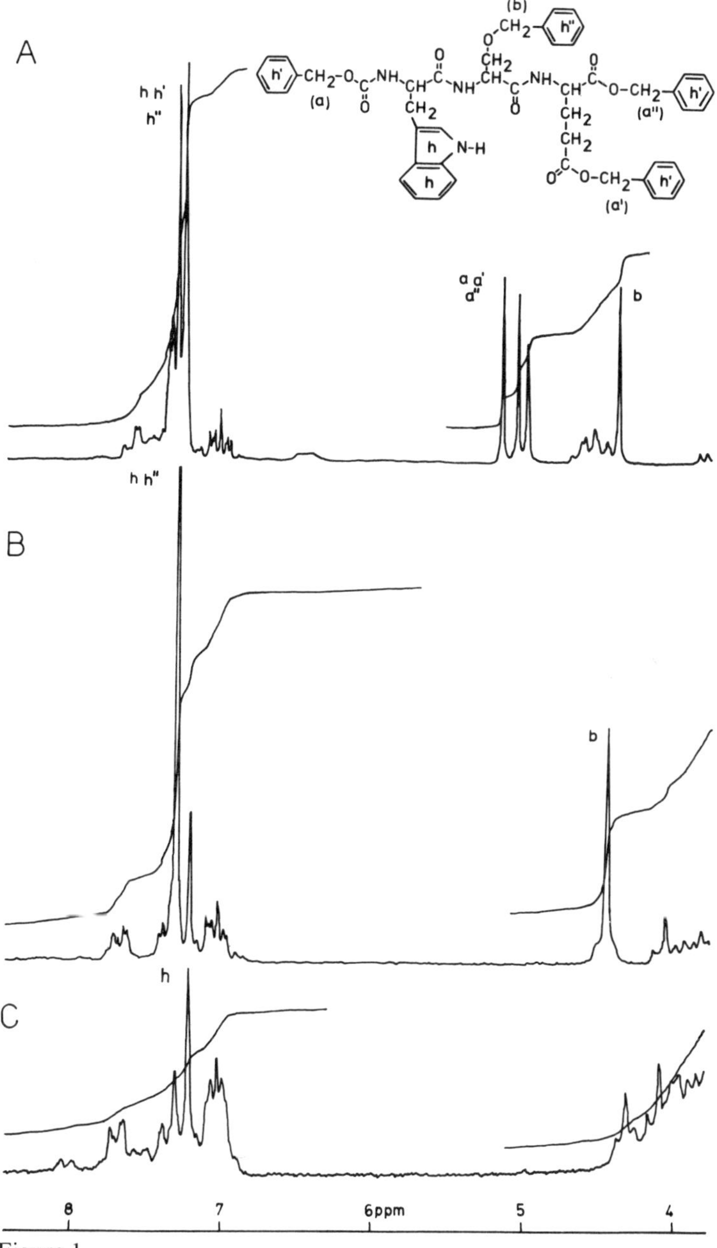

Figure 1
^{1}H-NMR-spectra (90 MHz, Standard TMS)
A: Z-Trp-Ser(BZL)-Glu(OBZL)-OBZL in $(CD_3)_2CO$
B: After partial hydrogenolysis, Trp-Ser(BZL)-Glu in $(CD_3)_2SO$ and D_2O
C: Trp-Ser-Glu in (CD_3) SO and D_2O

the 2–3 Hz delta frequencies. The quantification was achieved by automatic frequency analysis and expressed in mm deflexion per 10 sec. The activity during each 5-min period was directly measured from the curve of the wave analyzer by the operator himself and plotted diagrammatically: delta amplitude values in mm on the ordinate and time in periods of 5 min on the abscissa (Figure 2). The wave analyzer was calibrated by means of a signal generator (Wavetek Ltd, San Diego, USA) for 2 and 3 Hz, using calibration impulses of 50, 75, 100, 120 und 150 μV (amplitude). The mean integrated voltage of the calibration signal was calculated as the root mean square (RMS) of one-half its peak-to-peak value. The basic unit for both frequencies 2 and 3 Hz was 11.34 ± 2.11 μV $\times$ sec/1 mm height of the peak recorded. For conversion of mm deflexion values into absolute physical values, the data in mm measured by the operator were processed in a Univac 1108 computer system, which expressed the increased delta activities in RMS μV and their time integrals. The results of this elaborate computer analysis confirmed the validity of the automatic wave analysis and provided additional information which will be published later [25].

For assessment of its activity and comparison with the nonapeptide DSIP, the tripeptide was infused at the same concentration as DSIP, i.e. 25 nmol in 0.05 ml dialyzing solution [10] in 6 rabbit recipients.

Comparison of the effects of the tripeptide, of a control solution and of the Delta-Sleep-Inducing-Peptide (DSIP)

EEG effects. The intraventricular infusion of the tripeptide in rabbits did not significantly alter the delta activity (Fig. 2). This activity reached a value of 96% in the post-infusion period, as referred to the value of both reference periods taken as 100%. In a series of 13 rabbits, infused with dialyzing [10] control solution, the activity was 86% for the post-infusion period. The statistical analysis (standard error of the mean) confirmed that no significant difference exists between both curves during the post-infusion period. By contrast, the delta activity induced by infusion of natural Delta-Sleep-Inducing-Peptide, carried out under identical conditions in 4 rabbits, exhibited a delta increase of 138% against 92% in a series of 12 control rabbits (Fig. 3). This means that the delta activity developed by the natural nonapeptide is significantly greater than that of

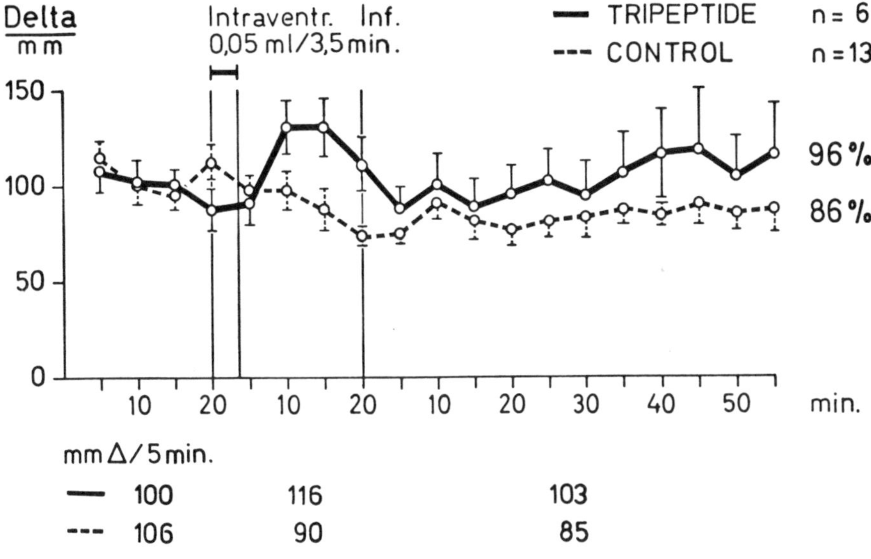

Figure 2
Variations of the EEG delta activity of the motor cortex after infusion of 25 nmol Trp-Ser-Glu and control solution.

the tripeptide (42%). From this comparison, it furthermore results that DSIP has a specific effect, since Trp-Ser-Glu and other oligopeptide analogues of different amino acid composition [25] cannot duplicate the hypnogenic effects.

Visceral effects. The heart rate exhibited, after infusion of tripeptide, during the post-infusion period, a value of 87% (decrease of 13%) against 90% after infusion of hypnogenic nonapeptide. In the control group, the heart rate decreased to 94%. After infusion of tripeptide, the respiration rate showed a decrease of less than 10%. This slight bradypneic effect, in absence of delta increase, should be considered as unspecific, in contrast to the bradypneic effect induced by the hypnogenic nonapeptide, characteristic of orthodox delta-sleep.
From all experimental data, it appears that the tripeptide induces no significant changes in EEG delta activity during the post-infusion period, and only a slight slowing of the heart and respiration rates.

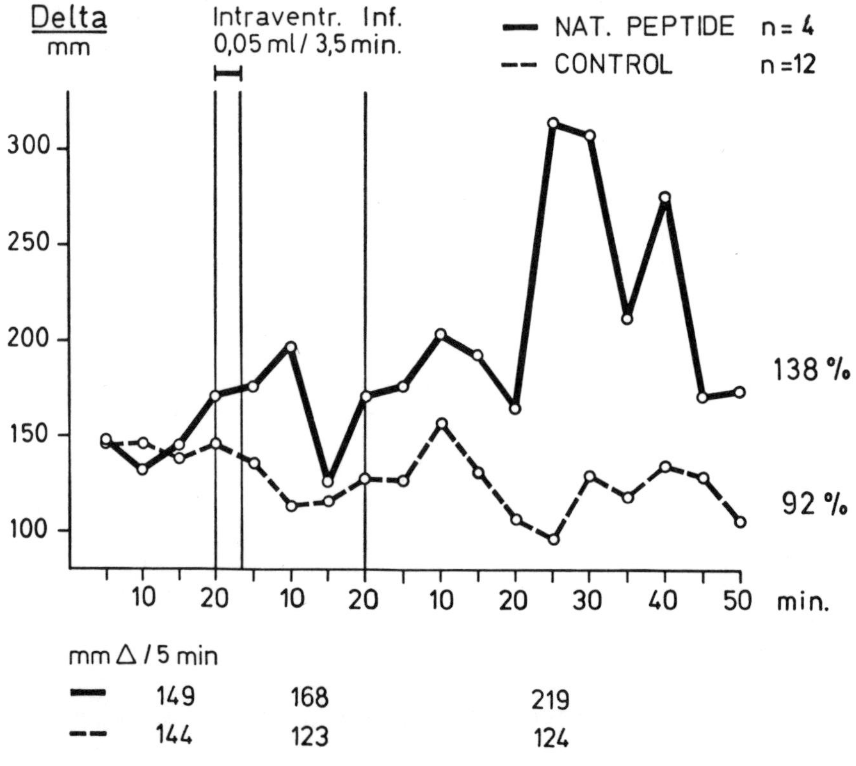

Figure 3
Variations of the EEG delta activity of the motor cortex after infusion of 25 nmol Delta-Sleep-Inducing-Peptide (DSIP) and control solution.

References

[1] R. Legendre and H. Piéron, *Des résultats histophysiologiques de l'injection intra-occipito-atlantoïdienne de liquides insomniques,* C. r. Soc. Biol. *1910,* 62. Séance du 25 juin.

[2] W.R. Hess, *Hirnreizversuche über den Mechanismus des Schlafes,* Arch. Psychiat. Nerv. Krankh. *86,* 287 (1927).

[3] M. Monnier, Th. Koller and S. Graber, *Humoral influences of induced sleep and arousal upon electrical brain activity of animals with crossed circulation,* Expl. Neurol. *8,* 263 (1963).

[4] E. Aserinsky and N. Kleitman, *Regularly occurring periods of eye motility, and concomitant phenomena, during sleep,* Science *118,* 273 (1953).

[5] M. Jouvet, *Etude de la dualité des états de sommeil et des mécanismes de la phase paradoxale,* in: *Aspects anatomofonctionnels de la physiologie du sommeil,* p. 397 (Ed. M. Jouvet; Centre National de la Recherche Scientifique, 1965).

[6] M. Monnier and L. Hösli, *Dialysis of sleep and waking factors in blood of the rabbit,* Science *146,* 796 (1964).

[7] J.R. Pappenheimer, G. Koski, V. Fencl, M.L. Karnovsky and J. Krueger, *Extraction of Sleep-Promoting Factor S from cerebrospinal fluid and from brain of sleep deprived animals,* J. Neurophys. *68,* 1299 (1975).

[8] H. Nagasaki, M. Iriki, S. Inoue and K. Uchizono, *The presence of a sleep-promoting material in the brain of sleep-deprived rats,* Proc. Japan Acad. *50,* 241 (1974).

[9] M. Monnier, A.M. Hatt, L.B. Cueni and G.A. Schoenenberger, *Humoral transmission of sleep. VI. Purification and assessment of a hypnogenic fraction of 'sleep dialysate' (Factor delta),* Pflügers Arch. *331,* 257–265 (1972).

[10] G.A. Schoenenberger, L.B. Cueni, M. Monnier and A.M. Hatt, *Humoral transmission of sleep. VII. Isolation and physical-chemical characterization of the 'Sleep inducing Factor Delta',* Pflügers Arch. *338,* 1 (1972).

[11] G.A. Schoenenberger and M. Monnier, *Isolation, partial characterization and activity of a humoral 'delta sleep' transmitting factor,* Brain and Sleep, p. 39–69 (De Erven Bohn BV 1974).

[12] J.E. Shields, W.H. McGregor and F.H. Carpenter, *Preparation of benzyl and p-nitrobenzyl esters of amino acids,* J. Org. Chem. *26,* 149 (1961).

[13] L. Zervas, M. Winnitz and J.P. Greenstein, *Esterification procedure for amino acids,* J. Org. Chem. *22,* 1515 (1957).

[14] G. Hillman, Z. Naturf. *1b,* 682 (1946).

[15] E.L. Smith, *Carbobenzoxy amino acids and peptides,* J. biol. Chem. *175,* 39 (1948).

[16] E. Wünsch and A. Zwick, *Zur Synthese des Glucagons I, Darstellung der Sequenz 1–4,* Chem. Ber. *97,* 2497 (1964).

[17] G. GNICHTEL and W. LAUTSCH, *Über die Synthese von Peptidderivaten des Porphyrins c,* Chem. Ber. *98,* 1647 (1963).

[18] H. SACHS and E. BRAND, *Benzyl ersters of Glutamic acid,* J. Am. chem. Soc. *75,* 4610 (1953).

[19] L.R. OVERBY and A.W. INGERSOLL, *Resolution of N-carbobenzoxy amino acids, alanin, phenylalanin and tryptophan,* J. Am. chem. Soc. *82,* 2067 (1960).

[20] J.S. FRUTON, *Synthesis of Peptides of* L*-Serin,* J. biol. Chem. *146,* 464 (1942).

[21] E. WÜNSCH, in: *Houben-Weyls Methoden der Organischen Chemie, Synthese von Peptiden,* vol. II (G. Thieme, Stuttgart 1974).

[22] V.M. MONNIER, *Der Nachweis der Abspaltung von Benzylschutzgruppen mittels ^{1}H-NMR-Spectroskopie,* Helv. chim. Acta *59,* 348 (1976).

[23] M. MONNIER and A.M. HATT, *Intraventricular Infusions in Acute and Chronic Rabbits,* Pflügers Arch. *317,* 268 (1970).

[24] G. FÖLSCH, *Synthesis of phosphopeptides, dipeptides, tripeptides and O-phosphorylated derivatives of* L*-serine,* Acta chem. scand. *20,* 459 (1966).

[25] G.A. SCHOENENBERGER and M. MONNIER, *Characterization of a delta EEG sleep inducing peptide (DSIP). Amino acid analysis, sequence, synthesis,* Proc. Nat. Acad. Sci. USA, vol. 74, in press (1977).

Abstract

In order to test the specificity of the natural Delta-Sleep-Inducing-Peptide (DSIP), a tripeptide with the same N-terminal amino acid was synthesized. The synthesis of the new tripeptide L-Trp-L-Ser-L-Glu was carried out by the method of the mixed anhydride. Protecting groups were all oxygen-bound benzyl groups. The physical-chemical data of the newly synthesized peptides are reported.
The biological activity of the tripeptide was assayed by intraventricular infusion in the rabbit under the same conditions as for the DSIP. The effects of the tripeptide on the EEG could not duplicate those of DSIP which induced a marked increase of delta activity, typical for orthodox 'Slow Wave Sleep' (SWS).

Experientia Supplementa

Gegenwartsprobleme der Ernährungsforschung – Present Problems in Nutrition Research – Les problèmes actuels de la nutrition*. Symposium Basel, 1.–4. X. 1952. Editor: F. Verzár (1953).
SBN 3-7643-0103-1

IVe Congrès international de chimie pure et ppliquée – XIV. Internationaler Kongreß für reine nd angewandte Chemie – XIVth International Congress of Pure and Applied Chemistry. Zürich, 1.–27. VII. 1955. 5 conférences principales et conférences de sections – 5 Hauptvorträge und Hauptsektionsvorträge – 5 Main Congress ectures and 9 Lectures in the Sections (1955).
SBN 3-7643-0104-X

ymposium über experimentelle Alternsforchung – Symposium on Experimental Research n Ageing. Basel, 4.–7. IV. 1956. Herausgeber/ ditor: Prof. Dr. F. Verzár (1956).
SBN 3-7643-0105-8

Ve Congrès international de chimie pure et ppliquée – XV. Internationaler Kongreß für eine und angewandte Chemie – XVth International Congress of Pure and Applied Chemistry. issabon, 8.–16. IX. 1956. 6 conférences princiales et 8 conférences de sections – 6 Hauptvorräge und 8 Hauptsektionsvorträge – 6 Main Congress Lectures and 8 Lectures in the Sections 1956).
SBN 3-7643-0106-6

5 Jahre hochalpine Forschungsstation Jungaujoch 1931–1956. Herausgegeben von Prof. r. A. von Muralt (1957).
SBN 3-7643-0107-4

VIe Congrès international de chimie pure et ppliquée – XVI. Internationaler Kongress für eine und angewandte Chemie – XVIth International Congress of Pure and Applied Chemistry. Paris, 18.–24. VII. 1957. 18 conférences lénières – 18 Main Lectures – 18 Hauptvoräge (1957).
SBN 3-7643-0108-2

ericht über die Tätigkeit der Schweizerischen tudienkommission für Atomenergie von 1946 is 1958 (1960).
SBN 3-7643-0109-0

ssays in Coordination Chemistry. Dedicated to erold Schwarzenbach. Edited by W. Schneider, . Anderegg and R. Gut (1964).
SBN 3-7643-0110-4

0
erz und Kreislauf der Säugetiere. Von J. Grauiler (1965).
SBN 3-7643-0111-2

1
eristeme. Von Otto Schüepp (1966).
SBN 3-7643-0112-0

2
roceedings of the 2nd International Symposium n Polarization Phenomena of Nucleons. arlsruhe, September 6–10, 1965. Editors: . Huber and H. Schopper (1966).
SBN 3-7643-0113-9

3
tmosphärische Spurenstoffe und ihre Bedeung für den Menschen. Symposium vom 18. und 9. Juni 1966 in der Klimaphysiologischen tation St. Moritz-Bad (Engadin) unter der eitung von Prof. Dr. Chr. Junge, Mainz (1967).
SBN 3-7643-0114-7

4
ymposium on New Trends in Basic Lymphology. ditors: J. M. Collette, G. Jantet, E. Schoffeniels 967).
SBN 3-7643-0115-5

15
Comparative Physiology of the Heart: Current Trends. Proceedings of a Symposium held at Hanover, New Hampshire (USA) 1968. Edited by F. V. McCann (1969).
ISBN 3-7643-0116-3

16
Methods for Transfecting Cells with Nucleic Acids of Animal Viruses: a Review. By G. R. Dubes (1971).
ISBN 3-7643-0574-6

17
The Challenge of Life. Biomedical Progress and Human Values. Roche Jubilee Symposium, August 31 to September 3, 1971 (1972).
ISBN 3-7643-0589-4

18
Biological Aspects of Electrochemistry. Proceedings of the 1st International Symposium Rome (Italy), May 31 to June 4, 1971. Edited by G. Milazzo, P. Jones and L. Rampazzo (1972).
ISBN 3-7643-0584-3

19
Untersuchungen zur Physiologie und Pathologie der Cervix uteri des Rindes. Von Dr. med. vet. H. F. Gloor (1973).
ISBN 3-7643-0662-9

20
New Concepts in Air Pollution Research. Interdisciplinary Contributions by an International Group of 20 young Scientifists. Edited by Jan-Olaf Willums (1974).
ISBN 3-7643-0681-5

21
Drogenmißbrauch und Gesetzgeber. Von Walter P. von Wartburg (1974).
ISBN 3-7643-0684-X

22
Contributions to Applied Statistics. – Dedicated to A. Linder (1976).
ISBN 3-7643-0721-8

23
Quantitative Structure Activity Relationships. By Milon Tichy (1976).
ISBN 3-7643-0744-7

24
Seventh International Conference on Cyclotrons and their Applications. Zürich, Switzerland, 19–22 August 1975. Editorial Chairman: W. Joho (1975).
ISBN 3-7643-0823-0

25
Proceedings of the 4th International Symposium on Polarization Phenomena in Nuclear Reactions. August 25–29, 1975 (1976).
ISBN 3-7643-0836-2

26
Enzymes and Proteins from Thermophilic Microorganisms. Structure and Functions. Proceedings of the International Symposium, Zürich, July 28 to August 1, 1975 (1976).
ISBN 3-7643-0827-3

27
Radioprotection. Chemical Compounds – Biological Means. A Symposium by Correspondence. Edited by A. Locker and K. Flemming (1977).
ISBN 3-7643-0871-0

28
Shallow-Water Sponges of Western Bahamas. By F. Wiedenmayer (1977).
ISBN 3-7643-0906-7

29
Synthesis of the Tripeptide L-Trp-L-Ser-L-Glu. Comparison of its biological activity with that of the Delta-Sleep-Inducing-Peptide (DSIP). By V. M. Monnier (1977).
ISBN 3-7643-0915-6 1977

Birkhäuser Verlag
Basel und Stuttgart

ISBN 3-7643-0915-6